MATHS

Simplify it is Sort out

Marcos Cervantes Janssen

First edition: October 15, 2022

Copyright © *2022 Marcos Cervantes Janssen*

Edited by Editorial letrRoja@

https://www.facebook.com/LETRA3ROJA

https://www.newtek.janssen@gmail.com

https://payhip. com/letra33roja

MATHEMATICS

https://newtekjanssen.es.tl/letra3roja@gmail.com

MATHEMATICS

is
Solving
By

: Marcos Cervantes Janssen

CONTENTS:

- PROLOGUE..................................5
- FORMULAS...................................7
- EQUATIONS.................................8
- VARIABLES..................................9
- CONSTANT................................10
- AVERAGES................................11
- TOLERANCE..............................12
- EVEN AND ODD: 13
- NATURAL NUMBERS................15
- PRIME NUMBER........................17
- IMAGINARY NUMBERS............20
- INFINITE NUMBERS..................23
- EPILOGUE.................................27

PROLOGUE:

It is said that mathematics is an exact science, Plus, a true mathematician knows what floating point means, negative numbers, and the world of fractions.

Likewise, simplifying and averaging are just tools to not get lost in this wonderful world of infinite results, therefore, mathematics will always be progressive, on the way to the punctual solution.

Mathematics is the solving force of real problems, through written numbers that accurately represent each movement of the problem to be solved.

Mathematics is the strokes and brushstrokes of a painting with infinite details, a painting of our existing and existential reality, thus a tool used since the beginning of our history.

Through words, ideas are captured and thus written, they are preserved for generations, so through numbers, forms and their existential reason endure in our culture, in the same way they can be studied more deeply, inherited to continue enjoying his unlimited knowledge, in our universe that is undoubtedly mathematical.

In this way, in this writing, a logical reasoning will be exposed by which the importance of numbers is manifested, which complete writing of the existential form, that in this way that all writing is exposed, as a result of thought and rational logic, as well Even grammar, like logic, entails laws and rules naturally conceived since forever.

It is discovering our mathematical nature which dazzles great possibilities of constant resolution.

FORMULAS:

Any instruction that defines a solution and that is duly written, efficiently systematizes the process, which can be replicated with the necessary precision for each matter.

This procedure with its components, which present a definite solution, can be simplified.

Already in their complex case, they can be studied for the understanding of the problem to be solved, so that the efficiency certainly does not depend on the degree of simplification, more so if its resolution frequency increases.

In this way, the most important thing is not the length but the precision through the inclusion of the largest number of unknowns, which will increase the solution and its expected result to be more efficient.

EQUATIONS:

Equations are a set of formulas, which contain a variety of unknowns, called variables, these are the ones that represent each of the parts of a problem or situation as the case may be.

This refers to different actions, which are equal, one from another, in this way that the name is derived, being called **EQUATIONS.**

The equation is not a solution formula for an isolated problem, the equation takes a multitude of unknowns, which are solutions of each other, thus we denote how each unknown is a co-participant of the solution as a whole, thus a solution is often shared by different problems, and an equation a system.

VARIABLES:

They are the elements, whose value is not defined at the moment, more through mathematics is when the value that corresponds to each variable is found, thus the equation, through the corresponding formulas and procedures, reveal the value of each variable, let us remember the importance of the variables as individual and important parts, for a completeness.

Each variable is by itself, important and unique, more when found in a system of inclusion of individuals, this is how each variable becomes an indispensable partner in the equation, finding the value of the variable is the punctual solution, the variable itself is a constant but in an unknown way, which requires a resolution process to reveal its true value in the equation.

CONSTANT:

A constant is a value determined by some stable phenomenon, this is very useful because it is an already known element, so the equation will have a beginning, a constant, which is why it is a fundamental basis for resolution in mathematics.

The constant is the opposite of the variable, the constant is given by nature and its already established laws, discovered throughout history by the human being; the constants are the definition of the way and form, which without any sudden change, bring stability, knowledge, to the structure. But as well as the functioning of our existence, an example of a constant is the number pi, also each natural number is a constant, because for example 3 will always be worth three, this in the situation that this, the constants appear in everything.

AVERAGE:

The average is a result of different situations, in an average, which gathers all the results values around one that represents them. As a common approximation, averaging is a solution of reconciliation between dangerous extremes.

The word average, means, to be in favor of the average, temperance is not the same as lukewarmness, all these concepts, although they seem to be mathematical, reveal the universal nature of mathematics in human life and existence, it is so that the The subject of mathematics concerns us in this writing, referencing the **whole** and not just numbers.

The average is the one who represents a large group of different units, is the one who measures the central tendency, with which a system that is too large and dispersed can have an identity to be known, analyzed and understood.

TOLERANCE:

Also called, margin of error, the lower the tolerance, the greater the accuracy or perfection, in the same way flexibility plays a role in tolerance, flexible systems have a sufficient percentage of tolerance, for compliance and no rupture of a specific process, this without giving rise to the dissolution or destruction in its entirety, due to uncontrolled fragmentation.

Tolerance is vital for solving problems in a faster way, because by having resolution movement margins, the possible solutions already in advance glimpse the way out, being that different solutions are already a parameter of advance to the last solution, it is so tolerance allows us to glimpse the solution in advance, each problem thus has a variety of paths for a single final answer.

EVEN AND ODD:

The numbers are divided firstly into two large groups, positive and negative, with even and odd being second, so in this way we have that an even number is symmetrical in its division, it is also that its division always gives as a result, integers, on the contrary the odd ones when divided in two give as results fractions, which contain in themselves, the so-called tolerance depending on the decimals they contain, that is how the even ones divided in two, and the odd ones take this mathematical property so important in the analysis that leads to the necessary operations in each situation, this as natural functions.

Odd numbers are so important and necessary because of their divisible variety and their balance in more than two parts, being balanced multilinks.

INTEGERS AND FRACTIONS:

"All integers are rational, that is, they can be expressed as fractions, although not all rational numbers are integers."

Rational numbers are represented as fractions and comprise all numbers that can be expressed as a division by two whole numbers.

On the other hand, fractions, as the name implies, are made up of an integer and a decimal part. Fractions can be represented in many ways: , with a tone plus .

The addition of integers and fractions is a mathematical operation that is performed to obtain the result of the addition of two or more numbers. This operation can be done manually or by using calculators.

NATURAL NUMBERS:

The nature of natural numbers is very interesting. They are often called "positive integers" since they only include positive integers. However, they also include zero. Therefore, they are sometimes called "positive integers and zero".

The nature of the natural numbers is very simple: they are all the numbers found in the sequence 1, 2, 3, 4, 5, 6, 7, 8, 9, 10, 11, 12, 13, 14, 15. .. and so on. As you can see, this sequence starts with the number 1 and has no upper limit; therefore, we can say that the natural numbers are all those that are in this sequence. from 1 onwards.

Natural numbers are so necessary to equations that they need to be included in almost every equation.

Natural Numbers: An Introduction Natural numbers are the positive integers, that is, the numbers 1, 2, 3, 4, 5, and so on. They can be used to count things or measure quantities. For example, we can count how many people are in a room using whole numbers. We can also measure the length of a table in meters or centimeters using natural numbers.

Natural numbers can be represented in various ways, for example, with drawings or symbols. In this document we are going to use symbols to represent the natural numbers. The most common symbols to represent natural numbers are the digits 0 to 9 (for example, 3 is represented as "3").

PRIME NUMBERS:

Prime numbers are those numbers that can only be divided by themselves and unity. That is, they cannot be divided by any other number. For example, the number 7 is a prime number, since it can only be divided by itself (7) and one (1). On the other hand, the number 6 is not a prime number, since it can be divided by 2 (3 times), 3 (2 times) and 6 (once). A PRIME NUMBERS on: . With a confident tone.

Prime numbers are numbers that can be divided by a single number. For example, 2 is a prime number because it can only be divided by one. Also, 0 is a prime number since it cannot be divided. Many famous things in mathematics and in the world have been made with prime numbers. For example, the number system of electronic circuits and digital clocks is based on prime numbers. Prime numbers are also essential for encryption, which is used to protect data in many cases.

The first prime number is known as 0 and it only has one factor: itself. Back then, people thought that 0 was the most basic element. In fact, the ancient Greeks sometimes called 0 for emptiness or the absence of something. Over time, people discovered that 0 is actually a number, and not just a letter, so it's interesting to see how our understanding of numbers changed. Since 0 was the first prime number, it is a symbol of principles and factors.

However, there are many factors that can be used to divide these numbers. Also, these numbers are very common: almost everyone knows at least three prime numbers. For example, 3 is a prime number because no number can be divided by exactly 3 without leaving a remainder. Therefore, 3 is an ideal prime number for many reasons, such as causing geometric shapes or being part of natural laws such as gravity.

The most common prime number is 3 and it is believed to be the number of God. The Catholic Church has its crosses in 3 on the beams of its sacred building. In addition there is the thirteen point fruit with 3 seeds in each of its forms and three seeds with 3 points each of its seeds. The multiplicity of the numbers 3 is believed to be the gift of the spirit for the reason that He said 'the Spirit emanates from the Word And the Word is Numbers.' Therefore, the multiplicity of the divine Spirit is the same as the Divinity itself, that is, the number 3.

IMAGINARY NUMBERS:

Imaginary numbers are a concept that is difficult to explain to others. A number is real if it exists in space and time, but imaginary if it doesn't. Imaginary numbers are part of mathematics, which is a way of thinking and communicating. Numbers are used in all fields of life and have many practical applications. For example, computers use numbers to perform calculations, and medical professionals use them to map human anatomy. Without imaginary numbers, the modern world would not work the way it does.

All numbers are imaginary, they are all based on infinity. 8 is the first imaginary number; it is called i and represents the number 1. Many more are added to create other numbers. The number 8 is represented by the letter 'i' because it looks like the capital letter 'I'. Imaginary numbers can be used to represent large amounts of data. They are especially useful when dealing with mathematical and scientific equations. Although not real, imaginary numbers have helped humanity immensely.

8 has special properties compared to the other seven imaginary numbers. It is positive and infinite. All other numbers are negative or finite, which means they come to an end. Also, the number 8 is even; all even numbers are positive and infinite too. Although they are not real, the 8 has many applications in the modern world.

Although imaginary numbers are not real, they are still a concern when working with mathematics. Working with 0 or 1 is not a problem since these do not exist in space or time. However, adding or subtracting imaginary numbers can be tricky. Adding 0+0=0 is easy since 0 exists in space and time. On the other hand, it is not possible to add an infinite number like 8 since infinity does not exist in space or time either. Therefore, it is best to continue working with real numbers when adding or subtracting imaginary numbers. Doing so won't compromise the results one bit, but it will make the process a lot easier.

Imaginary numbers are an integral part of mathematics and help mankind immensely without ever being real. Therefore, everyone must know how to use them. Numbers are everywhere in society; therefore, imaginaries are also necessary. Originality is key when creating new ideas; Without them, humanity would not be where it is today.

INFINITE NUMBERS:

Infinite number is an expression used to describe the infinite amounts of numbers. It was introduced by Georges Eugene Edouard Lemaître in 1918 as a response to Einstein's theory of relativity. According to Einstein, the number of infinite numbers is the same as the number of infinitesimal particles in the universe. In that way, the infinite number is a concept that illustrates the complexity of mathematics and the limits of human understanding.

Zero is a nameless number. It is represented by the letter 'x' and is used to initialize many number systems. For example, in astronomy, they have degrees, minutes, and seconds. In chemistry, you have moles, grams, and kilograms. In engineering, you have bolts and inches. The main use of zero is to simplify mathematical expressions and calculations. However, it is also used in financial transactions to keep track of savings accounts and bank balances.

As you can imagine, adding more zeros to a number makes it bigger, numerically speaking. Numbers with more zeros are called larger or higher infinite numbers. For example: 1,000,000 is an infinite number greater than 999,999 because the former has two zeros; 1,000,000 is a zeta plus a zeta; while the latter only has one zeta. The highest infinite numbers can go on forever because they can be expressed using different basic number systems. For example: 1,000,000,000 is expressed using the decimal system with ten as its base number system, this means it has 10 zeros (1 billion). The base number system for higher infinite numbers can also be infinite; this allows the Base Infinite Number System (BINS) to handle very large numbers.

Not all mathematicians agree on what the limit should be for higher infinite numbers; some say that there is not one. This is because, when considering larger basic number systems, such as hexadecimal (16) or octal (8), there are no limits to how large an infinite number can be. Furthermore, there is no limit when considering all possible natural numbers (from 0 to infinity). That means there is no limit to how many things there are in the universe, or how much information or knowledge we have about that information. Although our understanding of these infinite quantities is limited, we have still shown that mathematics is an essential tool used throughout society.

Infinite numbers are responses to the theories of cosmic proportions proposed by Albert Einstein in the early 1920s, who believed that space is made up of an infinite number of infinitesimal particles. Although we cannot understand infinity, mathematics continues to prove its value in everyday life. Infinite quantities help people to conceptualize and calculate large amounts of data and information. Although we are still discovering many applications for infinite numbers in our daily lives, they are still fascinating realms full of limitless possibilities, limitless!....

THE ETERNITY OF THE ABSOLUTE WHOLE

EPILOGUE:

Finally we find the great tool, which for centuries our civilization has evolved, so today we recognize the need to continue studying and deepening in all areas of mathematics, we should not think for a moment that the truth is within reach of our hands, because the universe reveals to us how immense and eternal it is His understanding, the path of mathematical understanding depends on the exercise, together with the practice of all its different resolution strategies.

We have more and better procedures that simplify the result, likewise the problems in each era are totally different, more with the same need to be resolved and thus evolve, practice always being the priority.

THE UNIVERSE IS MANIFESTED IN NUMBERS, ONLY FOR THE RATIONAL MIND.

All rights reserved. Under the sanctions established
in the legal system,
without written authorization from the holders of *Copyright* ©
the total or partial reproduction of this work by
any means or procedure
, reprography and computer processing

.

Hello, I am a researcher, writer and communications engineer, throughout my life, I have experienced strong situations in every way, I wish that your life goes better and better, and that you evolve as much as you can by expanding your knowledge, mind and will, I am sure we can find an expand our existence, I want to accompany you always, and I thank you in advance, you are.

It is said that mathematics is an exact science, Plus, a true mathematician knows what floating point means, negative numbers, and the world of fractions. Likewise, simplifying and averaging are just tools to not get lost in this wonderful world of infinite results, therefore, mathematics will always be progressive, on the way to the punctual solution. Mathematics is the solving force of real problems, through written numbers that accurately represent each movement of the problem to be solved. Mathematics is the strokes and brushstrokes of a painting with infinite details, a painting of our existing and existential reality, thus a tool used since the beginning of our history. Through words, ideas are captured and thus written, they are preserved for generations, so through numbers, forms and their existential reason endure in our culture, in the same way they can be studied more deeply, inherited to continue enjoying his unlimited knowledge, in our universe that is undoubtedly mathematical. In this way, in this writing, a logical reasoning will be exposed by which the importance of numbers is manifested, which complete writing of the existential form, that in this way that all writing is exposed, as a result of thought and rational logic, as well Even grammar, like logic, entails laws and rules naturally conceived since forever. It is discovering our mathematical nature which dazzles great possibilities of constant resolution.

www.ingramcontent.com/pod-product-compliance
Lightning Source LLC
Chambersburg PA
CBHW050327220526
45465CB00005B/2160